Weird World of Animals

Animals with Armor

By Mary Griffin

Gareth Stevens Publishing

Please visit our website, www.garethstevens.com. For a free color catalog of all our high-quality books, call toll free 1-800-542-2595 or fax 1-877-542-2596.

Library of Congress Cataloging-in-Publication Data

Names: Griffin, Mary, 1978- author
Title: Animals with armor / Mary Griffin.
Description: New York : Gareth Stevens Publishing, [2026] | Series: Weird world of animals | Includes bibliographical references and index.
Identifiers: LCCN 2025004271 (print) | LCCN 2025004272 (ebook) | ISBN 9781482471755 (library binding) | ISBN 9781482471748 (paperback) | ISBN 9781482471762 (ebook)
Subjects: LCSH: Armored animals–Juvenile literature
Classification: LCC QL940 .G75 2026 (print) | LCC QL940 (ebook) | DDC 591.47/7–dc23/eng/20250331
LC record available at https://lccn.loc.gov/2025004271
LC ebook record available at https://lccn.loc.gov/2025004272

Published in 2026 by
Gareth Stevens Publishing
2544 Clinton Street
Buffalo, NY 14224

Cover Designer: Nicholas Switalski
Interior Designer: Rachel Rising
Editor: Therese Shea

Photo credits: Cover, pp. 1, 16 Eric Isselee/Shutterstock.com; cover, pp. 1, 3-24 Jan_Vondrak/Shutterstock.com; p. 5 Vaclav Sebek/Shutterstock.com; p. 7 Ernie Cooper/Shutterstock.com; p. 8 Wirestock Creators/Shutterstock.com; p. 9 I Wayan Sumatika/Shutterstock.com; p. 11 ifish/iStock.com; p. 11 (inset) viper345/Shutterstock.com; pp. 13, 19 (inset) dwi putra stock/Shutterstock.com; p. 15 Paul Prints/Shutterstock.com; p. 17 Sophia C. Julian/Shutterstock.com; p. 19 CameraBaba/Shutterstock.com; p. 21 VANESSAL/Shutterstock.com.

Printed in the United States of America

CPSIA compliance information: Batch #CSGS26: For further information contact Gareth Stevens, at 1-800-542-2595.

CONTENTS

*Words in the glossary appear in **bold** type the first time they are used in the text*

AMAZING ADAPTATIONS

Adaptations are special body parts or ways of acting that help living things **survive**. Some adaptations are easy to see. For example, certain species, or kinds, of monkeys have tails that can grab tree branches. That helps the monkey get around in the treetops—and not fall! Other adaptations are less easy to see, but still important.

This book is about animals with body coverings that act like **armor**. That really comes in handy when they're faced with danger. Let's learn about this adaptation.

The skin of the rhinoceros is so thick that it's like armor, especially in places where it folds.

That's WEIRD!

A person's thickest skin is just 0.2 inch (4 mm) thick. Rhino skin can be as thick as 2 inches (5 cm)!

A Scorpion's Exoskeleton

Scorpions are **arthropods**. Like all arthropods, they have a hard outside covering. This covering, called an exoskeleton, protects their body's soft parts. They don't have bones in their body. Their **muscles** attach to their exoskeleton instead. So, the exoskeleton helps them move like our bones help us move. The exoskeleton is both tough and flexible, or bendable.

The exoskeleton of the stripe-tailed scorpion's tail is thicker than other kinds of scorpion tails. This adaptation helps it hunt prey, defend itself, and dig **burrows**.

The stripe-tailed scorpion is also called the devil scorpion. It can be found in the southwestern United States and northwestern Mexico.

That's WEIRD!

Scorpions have **pincers** and a tail that curves up over their body. Their tail ends in a stinger that contains poison!

As a scorpion's inner soft body grows, it gets too big for its exoskeleton. The old exoskeleton breaks apart and falls away. This is called molting. A new exoskeleton forms over the larger body.

The exoskeleton is mostly made of the **chemical** chitin (KY-tin). When the scorpion's body first **releases** chitin, the chitin is soft. It hardens over hours or days. The hardened exoskeleton protects scorpions from the bites and stings of many predators. Still, owls, lizards, snakes, **rodents**, and frogs have adaptations that help them eat scorpions.

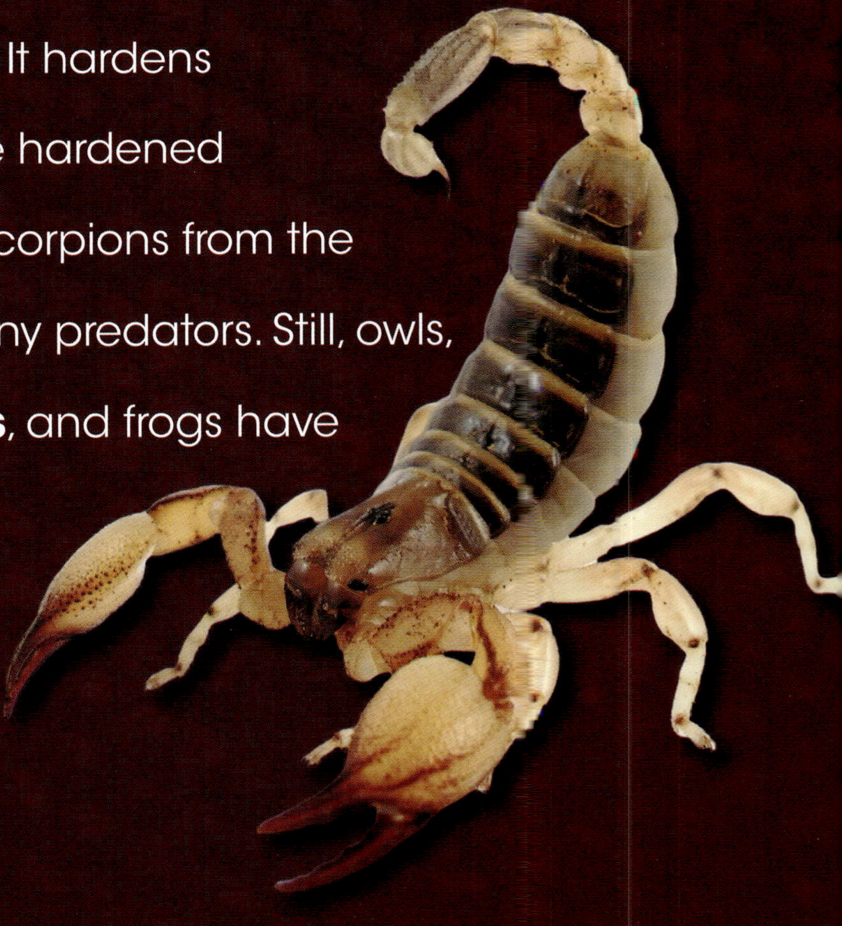

This scorpion is holding its prey with its pincers. It may have stung the prey. This could make the prey stop moving or even kill it.

That's WEIRD!

The exoskeleton at the end of a scorpion's pincers contains tiny bits of metal. That makes the pincers even tougher.

THE CREATIVE COCONUT CRAB

The largest arthropod on land is the coconut crab. It can be as large as 3 feet (0.9 m) wide! Coconut crabs live on islands of the South Pacific Ocean and the Indian Ocean.

Adult coconut crabs have a thick, hard exoskeleton. But they're not born like that. Juvenile, or young, coconut crabs' exoskeletons are thin. As they molt and grow, the exoskeleton gets thicker. Young coconut crabs need protection. They may wear an empty seashell or a coconut shell on their back!

Coconut crabs may find other things on the beach to wear for protection too!

That's WEIRD!

A coconut crab's pincers are so strong they can crack a coconut shell!

TOUGH TURTLE SHELLS

A turtle's shell is made of bone—it's part of its body. The top part, the carapace, is made of keratin. This is the same matter that makes up your fingernails. But it's much tougher than your nails! That's a good thing because it protects the turtle's **organs**.

Alligator snapping turtles are the biggest species of turtle that live in fresh water. They can weigh 175 pounds (79 kg)! They're known for the three lines of pointed **ridges** down the back of their carapace. It looks a bit like an alligator's back.

A turtle's shell is more than great armor! It keeps a turtle safe in many ways.

Terrific Turtle Shells

- keep turtles' inner bodies safe
- protect from predator attacks
- take in vitamin D from the sun
- help turtles swim faster
- color helps the turtle blend in with surroundings
- help turtles keep a healthy body temperature

That's WEIRD!

Snapping turtles have existed for about 90 million years. They lived alongside dinosaurs!

The Armadillo's Armor

The armadillo is a small **mammal** with short legs and a long tail. It has a pointy snout and long, strong claws for digging. There are about 20 species. The giant armadillo is about 5 feet (1.5 m) long and the pink fairy armadillo is just 6 inches (15 cm) long.

All armadillos have a bony body covering. This carapace is made of bands of hard keratin plates called scutes. They cover the body and head, and in many species, scutes cover the legs and tail too.

Most armadillos live in warm parts of Central and South America. The nine-banded armadillo, shown here, lives in the southern United States.

That's WEIRD!

Armadillos may stand on their back legs to sniff for predators around them. They can't see well.

Armadillos use their claws to dig in the ground and find bugs to eat. They also dig long burrows where they stay during the hot day—and hide from danger. Their predators include coyotes, bobcats, mountain lions, black bears, and alligators.

When faced with a predator, some run into their burrow. They expand their body so their armor sticks into the sides of their burrow. That makes it really hard for enemies to get them out. Other species of armadillos can tuck in their legs and roll into a ball—an armored ball!

When a three-banded armadillo rolls into a ball, all its soft parts are hidden.

That's WEIRD!

The nine-banded armadillo can't roll into a ball, but it can jump 3 feet (0.9 m) into the air! It also "plays dead."

The Pangolin's Scales

Pangolins are mammals that live in Africa or Asia. There are eight species. Unlike many mammals, pangolins have tough scales instead of hair over most of their body. Pangolin scales are made of **overlapping** keratin.

If a pangolin is scared, it rolls itself up. It covers its head with its scaly legs. It may whip around its tail, which has sharp scales, to defend itself against predators. These include leopards, hyenas, and large snakes called pythons.

Pangolins eat bugs and termites. They're sometimes called scaly anteaters, but they're not anteaters.

That's WEIRD!

Pangolins can protect themselves by releasing a bad-smelling liquid!

SO MANY MORE

Many more kinds of animals have a covering that acts as body armor. Millipedes are arthropods with exoskeletons. The pill millipede rolls into a ball when in danger. That makes it harder for a predator to eat it. Other animals with body armor include crabs, cockroaches, snails, and shellfish.

The hard outer coverings of these and many other animals give them a better chance at survival. They work so well that scientists study them to figure out how to make armor for people in dangerous places!

This photo shows a pill millipede rolled up for protection.

GLOSSARY

armor: The outer covering of an animal that keeps its body safe from harm. A thick covering worn to keep someone safe from harm.

arthropod: An animal that lacks a backbone and has a skeleton on the outside of its body, such as an insect, spider, shrimp, or crab.

burrow: A hole made by an animal in which it lives or hides

chemical: Matter that can be mixed with other matter to cause changes.

mammal: A warm-blooded animal that has a backbone and hair, breathes air, and feeds milk to its young.

muscle: One of the parts of the body that allow movement.

organ: A part inside an animal's body.

overlapping: Partly covering the same area.

pincer: A front claw on some animals that is used to hold things.

release: To set something free or let something go.

ridge: A narrow, raised part of a surface.

rodent: A small, furry animal with large front teeth, such as a mouse or rat.

survive: To stay alive.

FOR MORE INFORMATION

Books

Riggs, Kate. *Armadillos*. Mankato, MN: Creative Education and Creative Paperbacks, 2023.

Topacio, Francine. *What Do You Know About Animals with Armor?* New York, NY: PowerKids Press, 2021.

Websites

Alligator Snapping Turtle
nationalzoo.si.edu/animals/alligator-snapping-turtle
Read more about this turtle and watch a video.

Exoskeleton
kids.britannica.com/kids/article/exoskeleton/399448#
Learn about the hard outer body covering called an exoskeleton.

Publisher's note to educators and parents: Our editors have carefully reviewed these websites to ensure that they are suitable for students. Many websites change frequently, however, and we cannot guarantee that a site's future contents will continue to meet our high standards of quality and educational value. Be advised that students should be closely supervised whenever they access the internet.

INDEX